Earth & Space Science

—— Grade 6 ——

Written by Tracy Bellaire

The activities in this book have two intentions: to teach concepts related to earth and space science and to provide students the opportunity to apply necessary skills needed for mastery of science and technology curriculum objectives.

The experiments in this book fall under ten topics that relate to three aspects of **earth and space science**. In each section you will find teacher notes designed to provide you guidance with the learning intention, the success criteria, materials needed, a lesson outline, as well as provide some insight on what results to expect when the experiments are conducted. Suggestions for differentiation are also included so that all students can be successful in the learning environment.

Tracy Bellaire is an experienced teacher who continues to be involved in various levels of education in her role as Differentiated Learning Resource Teacher in an elementary school in Ontario. She enjoys creating educational materials for all types of learners, and providing tools for teachers to further develop their skill set in the classroom. She hopes that these lessons help all to discover their love of science!

Published in Canada by:
On The Mark Press
15 Dairy Avenue, Napanee, Ontario, K7R 1M4
www.onthemarkpress.com

OTM2157   ISBN: 9781770783638
© On The Mark Press

At A Glance

Learning Expectations	Our Solar System	The View from Earth	The Moon	Constellations	Asteroids, Comets, and Meteoroid	Space Exploration	Space Explorers	Going to the Extreme!	Technology in the Extreme	Aboriginal Contributions
Knowledge and Understanding Content										
Identify and describe the unique features of each planet in our solar system.	•									
Determine the positions of the sun, Earth, and the moon; describe the sun's position in the sky throughout a day.		•								
Determine the shape and position of the moon throughout a month.			•							
Identify the constellations in our night sky.				•						
Research and describe the unique features of asteroids, comets, and meteoroids.					•					
Research and describe the technology needed for space exploration, and the costs and benefits of this science.						•				
Research and explain contributions to space exploration, and how humans adapt to life in space.							•			
Identify extreme environments; research and present the exploration of an extreme environment.								•		
Research and explain technological advances of instruments used for extreme environment exploration.									•	
Research and explain the contributions of Aboriginal technologies to the exploration of extreme environments.										•
Thinking Skills and Investigation Process										
Make predictions, formulate questions, and plan an investigation.	•	•			•		•			
Gather and record observations and findings using drawings, tables, written descriptions.	•	•	•	•	•	•	•	•	•	•
Recognize and apply safety procedures in the classroom.	•	•	•	•	•	•	•	•	•	•
Communication										
Communicate the procedure and conclusions of investigations using demonstrations, drawings, and oral or written descriptions, with use of science and technology vocabulary.	•	•	•	•	•	•	•	•	•	•
Application of Knowledge and Skills to Society and the Environment										
Analyze the social and environmental costs and benefits of space exploration.						•				
Assess the contributions and benefits of space technology that allow humans to live in space.							•			
Evaluate responsible uses of technology that have led to important scientific discoveries vs. the possible misuses of technology that may affect society and our environment.									•	

OTM2157   ISBN: 9781770783638
© On The Mark Press

TABLE OF CONTENTS

Teacher Assessment Rubric

Student's Name: _____   Date: _____

Success Criteria	Level 1	Level 2	Level 3	Level 4
Knowledge and Understanding Content				
Demonstrate an understanding of the concepts, ideas, terminology definitions, procedures and the safe use of equipment and materials	Demonstrates limited knowledge and understanding of the content	Demonstrates some knowledge and understanding of the content	Demonstrates considerable knowledge and understanding of the content	Demonstrates thorough knowledge and understanding of the content
Thinking Skills and Investigation Process				
Develop hypothesis, formulate questions, select strategies, plan an investigation	Uses planning and critical thinking skills with limited effectiveness	Uses planning and critical thinking skills with some effectiveness	Uses planning and critical thinking skills with considerable effectiveness	Uses planning and critical thinking skills with a high degree of effectiveness
Gather and record data, and make observations, using safety equipment	Uses investigative processing skills with limited effectiveness	Uses investigative processing skills with some effectiveness	Uses investigative processing skills with considerable effectiveness	Uses investigative processing skills with a high degree of effectiveness
Communication				
Organize and communicate ideas and information in oral, visual, and/or written forms	Organizes and communicates ideas and information with limited effectiveness	Organizes and communicates ideas and information with some effectiveness	Organizes and communicates ideas and information with considerable effectiveness	Organizes and communicates ideas and information with a high degree of effectiveness
Use science and technology vocabulary in the communication of ideas and information	Uses vocabulary and terminology with limited effectiveness	Uses vocabulary and terminology with some effectiveness	Uses vocabulary and terminology with considerable effectiveness	Uses vocabulary and terminology with a high degree of effectiveness
Application of Knowledge and Skills to Society and Environment				
Apply knowledge and skills to make connections between science and technology to society and the environment	Makes connections with limited effectiveness	Makes connections with some effectiveness	Makes connections with considerable effectiveness	Makes connections with a high degree of effectiveness
Propose action plans to address problems relating to science and technology, society, and environment	Proposes action plans with limited effectiveness	Proposes action plans with some effectiveness	Proposes action plans with considerable effectiveness	Proposes action plans with a high degree of effectiveness

OTM2157   ISBN: 9781770783638
© On The Mark Press

Student Self-Assessment Rubric

Name: _____ Date: _____

Put a check mark ✔ in the box that best describes you.

Expectations	Always	Almost Always	Sometimes	Seldom
I listened to instructions.				
I was focused and stayed on task.				
I worked safely.				
My answers show thought, planning, and good effort.				
I reported the results of my experiment.				
I discussed the results of my experiment.				
I used science and technology vocabulary in my communication.				
I connected the material to my own life and the real world.				
I know what I need to improve.				

1. I liked _____

_____

2. I learned _____

_____

3. I want to learn more about _____

_____

OTM2157 ISBN: 9781770783638
© On The Mark Press

INTRODUCTION

The activities in this book have two intentions: to teach concepts related to earth and space science, and to provide students the opportunity to apply necessary skills needed for mastery of science and technology curriculum objectives.

Throughout the experiments, the scientific method is used. The scientific method is an investigative process which follows five steps to guide students to discover if evidence supports a hypothesis.

1. **Consider a question to investigate.**
   For each experiment, a question is provided for students to consider. For example, "What is out beyond our planet Earth?"

2. **Predict what you think will happen.**
   A hypothesis is an educated guess about the answer to the question being investigated. For example, "I believe that there are other planets, and there are other galaxies that could have planets that can support life forms". A group discussion is ideal at this point.

3. **Create a plan or procedure to investigate the hypothesis.**
   The plan will include a list of materials and a list of steps to follow. It forms the "experiment".

4. **Record all the observations of the investigation.**
   Results may be recorded in written, table, or picture form.

5. **Draw a conclusion.**
   Do the results support the hypothesis? Encourage students to share their conclusions with their classmates, or in a large group discussion format.

The experiments in this book fall under ten topics that relate to three aspects of earth and space science. In each section you will find teacher notes designed to provide you guidance with the learning intention, the success criteria, materials needed, a lesson outline, as well as provide some insight on what results to expect when the experiments are conducted. Suggestions for differentiation are also included so that all students can be successful in the learning environment.

ASSESSMENT AND EVALUATION:

Students can complete the Student Self-Assessment Rubric in order to determine their own strengths and areas for improvement. Assessment can be determined by observation of student participation in the investigation process. The classroom teacher can refer to the Teacher Assessment Rubric and complete it for each student to determine if the success criteria outlined in the lesson plan has been achieved. Determining an overall level of success for evaluation purposes can be done by viewing each student's rubric to see what level of achievement predominantly appears throughout the rubric.

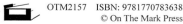

OTM2157   ISBN: 9781770783638
© On The Mark Press

OUR SOLAR SYSTEM

LEARNING INTENTION:
Students will learn about the unique features of each planet in our solar system.

SUCCESS CRITERIA:
- identify and describe unique features of the eight planets in our solar system
- participate in a quiz of knowledge about our solar system
- create a model of the solar system

MATERIALS NEEDED:
- a copy of "Our Solar System" Worksheet 1 for each student
- a copy of "The Planets" Worksheet 2, 3, 4, and 5 for each student
- a copy of "Planet Trivia" Worksheet 6, 7, 8, and 9 for each pair of students
- a copy of "A Model of the Solar System" Worksheet 10 for each student
- 9 Styrofoam balls (4 large, 4 small, 1 extra large), a piece of flat Styrofoam (15 cm x 15 cm), a large Styrofoam cup, 13 bamboo skewers (for each pair of students)
- assorted colors of paint and paint brushes, string, rulers, masking tape, scissors
- markers, pencils

PROCEDURE:
*This lesson can be done as one long lesson, or be divided into three shorter lessons.
1. Using Worksheet 1, 2, 3, 4, and 5, do a shared reading activity with the students. This will allow for reading practice and learning how to break down word parts in order to read the larger words in the text. Along with the content, discussion of certain vocabulary words would be of benefit for students to fully understand the passage.

Some **interesting** vocabulary words to focus **on** are:

- terrestrial
- dwarf
- surrounded
- erupt
- particles

- astronomy
- diameter
- surface
- rotate
- span

- atmosphere
- canyon
- horizontal
- hurricane
- orbit

2. Divide **students** into pairs. Give them Worksheets 6, 7, 8, and 9. They will cut out the trivia cards, creating two more questions to add to the game. Each student will take a turn asking their partner a planet trivia question (correct answer is highlighted).

3. Give pairs of students Worksheet 10 and the materials to create a model of the solar system. Read through the materials needed and what to do sections to ensure their understanding of the task before they begin. This project may span over a couple of days, as the painted planets will need time to dry before assembly can begin.

DIFFERENTIATION:
Slower learners may benefit by working as a small group with teacher support in order to participate in the planet trivia game. They may need to reference the information as they provide answers to the questions. This would allow these learners to have another opportunity to re-read the information on Worksheets 1, 2, 3, 4, and 5 in a small group.

For enrichment, faster learners could choose a planet and **create** a travel brochure to convince people to **visit it**. (Provide sample brochures so students **understand** the purpose and layout). Students' **travel** brochures should include:

- location, size, composition, appearance (unique features), temperature, weather, moons, length of day and year, how long it takes to travel there from Earth, its distance from the sun.

Name:

Our Solar System

Have you ever wondered what is out there, beyond Earth?

Earth is just one of **eight planets** and **five dwarf planets** that orbit the Sun, along with many other objects such as asteroids, comets, meteoroids, and satellites.

The four planets closest to the sun are Mercury, Venus, Earth, and Mars. They are called the **terrestrial planets** because they have solid rocky surfaces.

The four large planets, Jupiter, Saturn, Uranus, and Neptune, are called the **gas giants** because they have a thick atmosphere of hydrogen and helium.

Did You Know?

Pluto was the ninth planet in our solar system until 2006 when it was demoted to a **dwarf planet**. A professor discovered an icy object about the same size as Pluto, out beyond its orbit. This object is the dwarf planet called Eris. It turns out that Pluto is just a fraction of the icy objects in its realm, so it is now considered a dwarf planet.

OTM2157   ISBN: 9781770783638

The Planets

Mercury is the closest planet to the sun. Its diameter is 4880 kilometres. It can get as hot as 430 °C or as cold as -170 °C. It has no weather patterns or water. On its surface there are mountains and craters, and it has a heavy iron core.

It takes Mercury 88 days to go around the sun once. This means that one year on Mercury is equal to 88 Earth days.

Venus is the second planet from the sun. It is 12 104 kilometres in diameter. It is surrounded by clouds containing a poisonous gas called sulfuric acid. These clouds trap the sun's heat. As a result, the temperature on Venus is about 450 °C. This hot temperature causes it to glow red.

It takes Venus 225 days to orbit the sun. This means one year on Venus is equal to 225 Earth days.

The surface of Venus is rocky and desert-like, hot and dry. There is no water. Much of its surface is covered by hardened lava flows from volcanoes that erupted a long time ago.

Did you know that Venus spins backwards, in the opposite direction that other planets do?

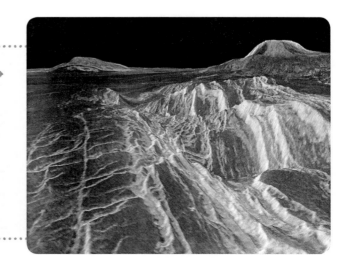

OTM2157   ISBN: 9781770783638
© On The Mark Press

9

The **Earth** is the third planet from the sun. Its diameter is 12 756 kilometres. The Earth has hot deserts and cold, icy poles. The average temperature on Earth is 22 °C, which makes it ideal for life forms. It is the only planet known to have life on it. The Earth's atmosphere contains mainly oxygen and carbon dioxide, and it protects life forms from the rays of the sun. It takes 365 days for the planet Earth to rotate around the sun. This is equal to one Earth year.

Mars is the fourth planet from the sun. It is 6 787 kilometres in diameter. It has two moons, Phobos and Deimos. Its temperature ranges from -120 °C during the winter and 25 °C in the summer. It has a thin atmosphere made mostly of carbon dioxide.

Mars is nicknamed the **Red Planet** because of the red color caused by rusted iron in its dusty, rocky surface. Mars has big volcanoes, and many channels and canyons on its surface which could have been eroded by water a long time ago. Presently, there is no liquid water on Mars, and instead only frozen water at its polar ice caps. A year on Mars is as long as 687 Earth days.

Almost 3/4 of the Earth is covered by water. The rest of Earth is land. The land and water are above the Earth's crust.

Under the Earth's crust is a deep layer of rock called the mantle. Under this is the core, which is made up of mostly iron and nickel.

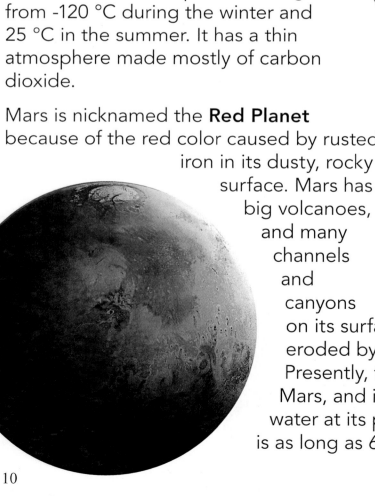

OTM2157   ISBN: 9781770783638

Jupiter is the fifth planet from the sun. The diameter of this huge planet is 142 200 kilometres. Jupiter is a cold planet, with the average temperature being -150 °C.

Its atmosphere and surface are mostly made up of hydrogen and helium. Jupiter has a great red spot, which is a giant mass of swirling gases. This spot is like a hurricane storm that has been raging for hundreds of years. One Earth year takes almost twelve Jupiter years.

Jupiter has 50 **known** moons circling around it. It also has three **rings**, made of fine particles like dust, that are more easily seen when backlit by the sun.

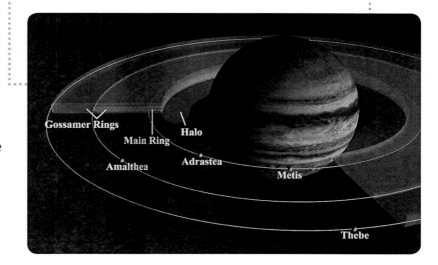

Gossamer Rings   Halo   Main Ring   Amalthea   Adrastea   Metis   Thebe

Saturn is the sixth planet from the sun. This large ringed planet has a diameter of 119 300 kilometres. It is a cold planet, with the average temperature being -180 °C.

Its atmosphere and surface are mostly made up of the gases hydrogen and helium. Saturn's ring system is actually made of icy objects flying around the planet. This ring system spans out hundreds of thousands of kilometers from the planet. Saturn has 53 known moons within its magnetic field. It takes Saturn 29.5 years to orbit the sun one time.

OTM2157   ISBN: 9781770783638

Uranus is the seventh planet from the sun. It is 51 800 kilometres in diameter. It is a cold planet, with the average temperature being -215 °C. Its atmosphere and surface are mostly made up of the gases hydrogen, helium, and methane. The methane gas is what gives Uranus its blue-green color.

Uranus has 27 known moons in its orbit. It has a system of 11 rings around it. It takes Uranus 84 years to orbit the sun once. Uranus rotates on an axis that is nearly horizontal, as though it has been knocked on its side. Due to this, its seasons last over 20 years long.

Neptune is the eighth planet from the sun. It is 30 800 kilometers in diameter. It is a cold planet with the average temperature being -220 °C.

Its atmosphere and surface are mostly made up of the gases hydrogen, helium, and methane.

Neptune has visible and very active weather patterns. This planet has fast moving winds of about 2000 km per hour! These fast moving winds are responsible for creating the great dark spot on this planet.

Neptune has 13 known moons, and a system of 6 rings that are very faint. It takes Neptune 165 years to orbit the sun one time. The long orbital period of Neptune means that the seasons on this planet last for forty Earth years!

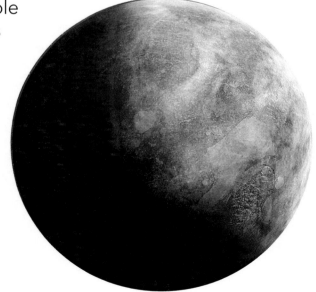

12

OTM2157   ISBN: 9781770783638

Name:

Planet Trivia!

Cut out the trivia cards. Place in a pile face down. Take turns reading aloud with a partner. Each correct answer gets a point.

Which planet is mainly covered with water?

    A. Mars

    B. Earth

    C. Neptune

What is the object that all planets orbit around?

    A. The sun

    B. The moon

    C. Mercury

Which planet is nicknamed the Red Planet?

    A. Jupiter

    B. Mars

    C. Venus

Which planet has a great red spot?

    A. Mars

    B. Venus

    C. Jupiter

Which planet is the hottest?

    A. Venus

    B. Mars

    C. Mercury

How many terrestrial planets are there?

    A. 4

    B. 5

    C. 6

Which planet has no atmosphere?

    A. Neptune

    B. Mars

    C. Mercury

What planet has the widest temperature range?

    A. Earth

    B. Mercury

    C. Saturn

OTM2157 ISBN: 9781770783638
© On The Mark Press

Name:

More Planet Trivia!

Cut the trivia cards out on the dotted lines. Place in the pile face down. Take turns asking questions to your partner.

Which planet has the only known life forms?

    A. Mars

    B. Jupiter

    C. Earth

Which planet is surrounded by clouds containing a poisonous gas called sulfuric acid?

    A. Venus

    B. Neptune

    C. Uranus

How many known planets are in our solar system?

    A. 9

    B. 8

    C. 10

How many known dwarf planets are in our solar system?

    A. 5

    B. 6

    C. 4

What planet is closest to the sun?

    A. Mars

    B. Venus

    C. Mercury

What planet is furthest from the sun?

    A. Saturn

    B. Neptune

    C. Uranus

How many days does it take Earth to orbit the sun?

    A. 12

    B. 225

    C. 365

What planet has rusted iron in its dusty, rocky surface?

    A. Venus

    B. Mars

    C. Jupiter

OTM2157   ISBN: 9781770783638
© On The Mark Press

Name:

More Planet Trivia!

Cut out the trivia cards. Place in a pile face down. Take turns reading aloud with a partner. Each correct answer gets a point.

Phobos and Deimos are moons of which planet?

    A. Neptune

    B. Mercury

    C. Mars

The layer beneath the Earth's crust is called the _____

    A. inner core

    B. mantle

    C. outer core

I rotate on an axis that is nearly horizontal. Which planet am I?

    A. Saturn

    B. Uranus

    C. Neptune

Which planet takes about 165 years to orbit the sun?

    A. Neptune

    B. Jupiter

    C. Venus

I rotate around the sun about every 88 days. Which planet am I?

    A. Mars

    B. Venus

    C. Mercury

I spin in the opposite direction that other planets do. Which planet am I?

    A. Saturn

    B. Earth

    C. Venus

Which planet has seasons that last for about 40 years long?

    A. Jupiter

    B. Neptune

    C. Uranus

I am the third planet from the sun. Which planet am I?

    A. Earth

    B. Venus

    C. Mars

OTM2157   ISBN: 9781770783638
© On The Mark Press

Even More Planet Trivia!

Cut these last trivia cards out on the dotted line. Place in the pile face down. Each partner will create one to add to the game.

Which planet's surface is covered by hardened lava flows from volcanoes that erupted long ago?

A. Mars

B. Mercury

C. Venus

Which planet is the largest?

A. Jupiter

B. Saturn

C. Uranus

Which planet is the smallest?

A. Venus

B. Earth

C. Mercury

Which planet is the closest in size to Earth?

A. Mars

B. Venus

C. Uranus

Which planet has seasons that last over 20 years long?

A. Uranus

B. Neptune

C. Saturn

Which planet has moving wind speeds of about 2000km per hour?

A. Jupiter

B. Neptune

C. Uranus

OTM2157 ISBN: 9781770783638
© On The Mark Press

Name:

A Model of the Solar System

You have learned a lot of interesting facts about the planets. Now it is time to create a model of our solar system!

Materials needed:

- 9 Styrofoam balls
  (4 large, 4 small, 1 extra large)
- assorted colors of paint and paint brushes
- a piece of flat Styrofoam (15 cm x 15 cm)
- a large Styrofoam cup
- a piece of string (30 cm long)

- 13 bamboo skewers
- a ruler
- scissors
- a marker
- masking tape

What to do:

1. Paint the Styrofoam balls similar in color as they are shown in the pictures on Worksheets 2, 3, 4, and 5. (The 4 small will be Mercury, Venus, Earth, and Mars. The 4 large will be Jupiter, Saturn, Uranus, and Neptune. The biggest one is the sun.)

2. Cut the bottom out of the large Styrofoam cup so that the "sun" can sit upright in it when the cup is turned upside down on a table.

3. Using the piece of string, place it around "Saturn". Mark the place on the string where it meets the end. Cut it to the length you need. (This will help you to make the rings).

4. Take the string you cut and make a circle with it on the flat piece of Styrofoam. Carefully outline it with the marker. Then draw **another ring** on the outside of this (about 2 cm away from the inner ring). Cut out the **inner circle**. Then cut out along **the outside of the outer ring**.

5. Once the planets are dry, use the skewers to connect them to the sun. To make them different distances from the sun, you may need to cut or break off pieces of some of the skewers.

6. Use pieces of skewers, masking tape, and a marker to create flag-like labels for the planets.

THE VIEW FROM EARTH

LEARNING INTENTION:

Students will learn about the positions of the sun, Earth, and the moon; about the sun's position in the sky throughout a day.

SUCCESS CRITERIA:

- describe how the Earth rotates on its axis and how it orbits the sun
- describe how the moon rotates and orbits the Earth, moving as one unit orbiting the sun
- demonstrate the rotation and orbit of the Earth and its moon around the sun
- create a sundial
- describe how the sun changes position throughout the day

MATERIALS NEEDED:

- a copy of "Day and Night" Worksheet 1 for each student
- a copy of "Revolving Around the Sun" Worksheet 2 for each student
- a copy of "A Look at the Moon" Worksheet 3 for each student
- a copy of "Rotate and Orbit!" Worksheet 4 and 5 for each student
- a copy of "Let's Create a Sundial!" Worksheet 6, 7, and 8 for each student
- a large ball, a small ball, a hat, 2 index cards (for each group of 3 students)
- a globe
- a large piece of poster paper or white Bristol board, a large empty can, a long stick or piece of dowelling, sand or small rocks, a ruler (for each pair or group of students)
- a sunny day
- a clock
- masking tape, markers, pencils

PROCEDURE:

*This lesson can be done as one long lesson, or be divided into three shorter lessons.

1. Using Worksheet 1, 2, and 3, do a shared reading activity with the students. This will allow for reading practice and learning how to break down word parts in order to read the larger words in the text. Along with the content, discussion of certain vocabulary words would be of benefit for students to fully understand the passage.

Some interesting vocabulary words to focus on are:

- axis
- orbit
- revolution
- rotate
- tilted
- hemisphere
- exposure
- journey
- equator

(As the moon orbits the Earth, the phases change. A phase of the moon is how much of the moon appears to us on Earth to be lit up by the sun. As the moon begins to orbit the Earth, we can only see a portion of the lit up side. As it continues to orbit the Earth, we see more of the lit up side until finally the moon is on the opposite side of the Earth from the sun and we get a full moon that is 100% lit up. As the moon continues orbiting, we see less of the lit up side. When we can't see any of the lit up side, this is called a new moon. In the new moon phase, the moon is between Earth and the sun.)

2. *This activity can be done in groups, or as a whole group demonstration. Students will **need teacher guidance** to demonstrate the Earth's orbit around the sun correctly.

   Give them Worksheets 4 and 5 and the materials to create a model of the positioning of the sun, the Earth, and its moon in the solar system. Read through the materials needed and what to do sections to ensure their understanding of the task before they begin. At this point, use the globe to show students how the Earth is divided into two

OTM2157   ISBN: 9781770783638
© On The Mark Press

hemispheres, Northern Hemisphere and Southern Hemisphere. Pointing out where the students live on the globe would be of interest to them also.

Once students have positioned themselves, the teacher will help the "Earth" to orbit the sun correctly. This means that as the "Earth" is approaching the half way around mark, the teacher ensures that the "Earth" begins to face away from the sun, so that the Northern Hemisphere now becomes further from the sun while the Southern Hemisphere becomes closer to the sun.

3. Explain to students that they will investigate how the sun moves in the sky through the day, and how it can create shadows. They will investigate this idea by creating their own sundials. This activity can be done in pairs. Give students Worksheet 6 and read through the materials needed and what to do sections to ensure students understand. Give students the materials and Worksheets 7 and 8. They will conduct the investigation, record observations, and make conclusions. *An option at the end of the activity is to have groups share their observations with another group to compare their findings.*

(Upon completion of this activity, students should understand that shadows are cast by objects that block the path of sunlight. In the morning and late afternoon, when the sun is lower in the sky, shadows are longer because the line from the sun to the object to the ground is flatter, making a longer shadow. Around noon, when the sun is directly overhead, there is almost no sunlight being blocked from hitting the ground, so a very small shadow is created.)

DIFFERENTIATION:

Slower learners may benefit by having another opportunity **to** re-read the information on Worksheets 1, 2, and 3 in a small group with teacher support. An additional accommodation would be to **have** these learners work in a small group **with** teacher support to complete the conclusion section on Worksheet 8. This would allow **for** an opportunity to discuss their observations **before** reaching conclusions.

For enrichment, faster learners could access the internet **to** research some interesting facts about the sun, **the** Earth, or the moon. This could be later shared with the large group to promote discussion **on** the topic.

Day and Night

You have learned that the planet Earth is third in line from the sun. In measured distance, it is about 150 million kilometres from the sun. Even from that great distance, the sun is Earth's powerful heat energy source.

So what is the Earth doing while it is soaking up the sun's heat energy? It is rotating! The Earth has an axis that runs from the North Pole to the South Pole.

As the Earth spins on its axis, from east to west, we experience day time and night time. The Earth takes 24 hours to complete a rotation.

Its rotation gives each part of the Earth a turn to be warmed by the sun. Life forms on Earth need the heat and light from the sun. If the Earth did not rotate, one half of the Earth would always be too hot to support life, and the other half would be in a deep freeze!

Axial Tilt of the Earth

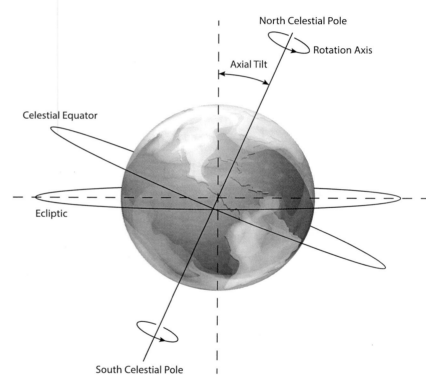

How long does it take for the sun's heat to reach Earth? Let's do the math!

We are 150 million km away from the sun. Light travels 300 000 km per second. Dividing these numbers, it will equal to 500 seconds, or 8 minutes and 20 seconds.

20

OTM2157 ISBN: 9781770783638
© On The Mark Press

Revolving Around the Sun

At the same time that the Earth is rotating on **its** axis, it is taking a journey around the sun. This journey is called **a re**volution.

The Earth's journey around the sun is about 940 million kilometres long and it takes one year (365 days) to complete.

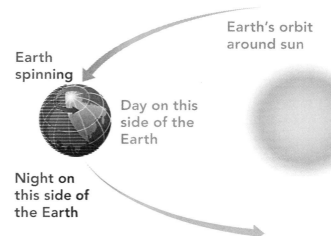

| Did You Know? |

The Earth is divided into two hemispheres. They are the Northern Hemisphere and the Southern Hemisphere. As the Earth orbits the sun during a year, its tilt causes the Northern **and** Southern hemispheres to change from more or less exposure to the **sun**.

As a hemisphere is tilted away from the sun, **the** length of a day gets shorter and it gets colder. As it is tilted towar**d the** sun, day lengths get longer and it gets warmer. The further yo**u are** from the equator in either hemisphere, the more obvious the effect. This is why we have the different seasons in the year, Spring, Summer, Winter, and Fall.

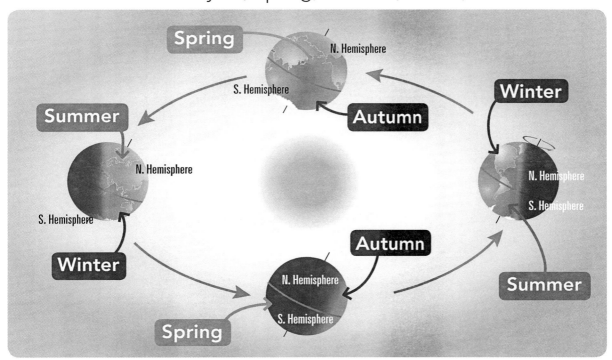

OTM2157 ISBN: 9781770783638
© On The Mark Press

A Look at the Moon

Are you wondering what the moon is doing while the Earth is revolving around the sun? It is orbiting the sun with the Earth as one unit. While it is on this journey, the moon is also revolving around the Earth, and rotating.

Have you ever wondered why the moon looks different in the sky some nights? The moon does not shine light, by itself. What we see when we look at the moon from Earth is the sun's light bouncing off the moon.

Sometimes we only see part of the moon, sometimes we see the full moon, and other times we can't see the moon at all.

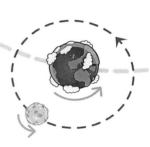

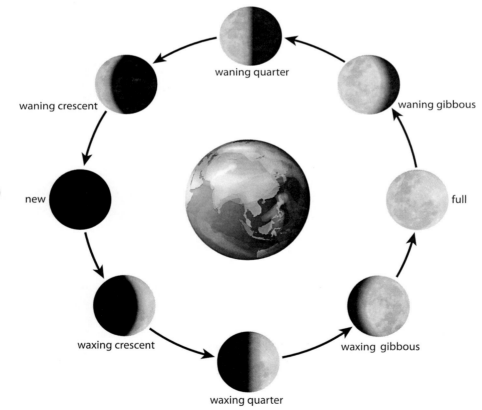

It takes about 28 days for the moon to orbit the Earth and to rotate.

The moon goes through many phases as it rotates and orbits the Earth, but from Earth we always see the same side of the moon.

OTM2157   ISBN: 9781770783638
© On The Mark Press

Rotate and Orbit!

It is time to review the facts. You have learned that:

- Earth rotates on its axis while it orbits the sun
- because of Earth's tilt, one hemisphere is closer to the sun for half the year, while the other is farther away
- the moon rotates as it orbits the Earth
- the moon orbits the sun as one unit with the Earth

Now it is time to put these ideas into practice!

Materials needed:

- a large, and a small ball
- a card with an **N** on it
- a card with an **S** on it
- a hat
- masking tape
- 2 other group members

What to do:

1. Tape the letter **N** to the hat. (**N** means Northern Hemisphere). One partner will wear the hat.

2. Tape the **S** to the lower back of this person. (**S** means Southern Hemisphere). This person is the Earth.

3. Another group member will hold the large ball (the sun).

4. The Earth will stand about 2 metres away from the sun, facing it, tilted forward a little.

5. As your teacher is turning the Earth around the sun, observe which hemisphere becomes closest to the sun. Record your observations on Worksheet 5.

6. The third group member will hold the small ball (the moon). The moon will rotate slowly while it orbits the Earth. Keep up with the Earth's orbit!

7. Make a conclusion about the rotation and orbit on Worksheet 5.

OTM2157   ISBN: 9781770783638
© On The Mark Press 23

Let's Observe

Choose words from the Word Box to complete your observations.

Northern	higher	lower

When the Earth **started** to turn, the _____

Hemisphere was closer to the sun, and the sun stayed

_____ in the sky for a longer time.

When the Earth was **half way around**, the _____

Hemisphere was farther from the sun, and the sun stayed

_____ in the sky for a longer time.

Illustrate your observations.

Let's Conclude

Choose words from the Word Box to complete your conclusions.

summer	winter	shorter	longer

When the sun is higher in the sky, the days are _____ ,

just like in the _____.

When the sun is lower in the sky, the days are, _____ ,

just like in the _____.

OTM2157   ISBN: 9781770783638
© On The Mark Press

Let's Create a Sundial!

You have learned that the length of day on Earth depends on the position of its rotation around the sun, bringing us through the different seasons of the year. As the Earth rotates on its own axis in a 24 hour time period, the sun appears to us as sitting at a different spot in the sky at different times of the day.

Did you know that the sun can be tracked in the sky using shadows? Let's explore this idea by creating a sundial to track the sun!

Materials needed:

- a large piece of poster paper
- a clock to tell the time
- a large empty can
- sand or small rocks
- a marker
- a ruler
- a long stick
- a sunny day!

What to do:

1. In the morning, place a large piece of poster paper on the ground where the sun will shine on it all day.

2. Put the metre stick inside the can and fill it with sand or rocks so that the stick stands straight up.

3. Using the marker and a ruler, draw a line along the shadow that is made by the stick.

4. Write down the time next to the line that you drew.

5. Repeat steps 3 and 4 at each hour of the day.

6. Record your observations on Worksheet 7.

7. On Worksheet 8, make conclusions about the movement of the sun and the shadows it created.

OTM2157   ISBN: 9781770783638
© On The Mark Press

Name:

Let's Observe

Illustrate what your sundial looked like at the end of the day.

What shape was the shadow?

_____

_____

When did the longest shadows happen?

_____

_____

When did the shortest shadows happen?

_____

_____

OTM2157   ISBN: 9781770783638
© On The Mark Press

Let's Conclude

What made the shadows?

_____

_____

_____

_____

Why did the length of shadows change?

_____

_____

_____

_____

Challenge question:

Do you think the same shadows will occur at the same times tomorrow?
Explain why or why not.

THE MOON

LEARNING INTENTION:

Students will learn about the shape and position of the moon throughout a month.

SUCCESS CRITERIA:

- identify the different phases of the moon
- track the change in shape and position of the moon during one month
- explain the changes that are seen in the moon during one month
- describe the effects of a lunar and a solar eclipse
- create a mobile to show the different phases of the moon

MATERIALS NEEDED:

- a copy of "The Phases of the Moon" Worksheets 1, 2, 3, and 4 for each student
- a copy of "Let's Go on a Moon Watch!" Worksheets 5 and 6 for each student
- a copy of "Eclipsed!" Worksheet 7 for each student
- a copy of "Making a Mobile of the Moon" Worksheet 8 for each student
- a copy of "Moon Phase Name Cards" Worksheet 9 for each student (on card stock)
- **2 copies** of "Moon Phase Creation Cards" Worksheet 10 for each student (on card stock)
- a model of the Sun, Earth, and Moon (that is able to show rotation)
- black pencil crayons (one per student)
- chart paper, markers, pencils, scissors, yarn
- coat hangers (one per student), or 2 sticks and string (per student)
- access to the internet

PROCEDURE:

*This lesson can be done as one long lesson, or be divided into five shorter lessons. Items 1, 2, 4, 5, and 6 can be done at school. Item 3 contains a homework component.

1. Explain to students that the moon moves around the Earth, taking about one month to go fully around the Earth. Give them Worksheet 1, read it through and discuss the moon's rotation. Divide students into pairs and give them Worksheets 2 and 3. Allow them time to do a "think-pair-share" activity to brainstorm some possible answers to why the moon looks different as it rotates around the Earth.

2. Come back as a large group. Have a discussion about the moon's rotation by asking some pairs to share their answers. *An option is to record student responses on chart paper.* Then using a model of the Sun, Earth, and Moon, demonstrate for students how the moon rotates. Explain why it looks different throughout its rotation around the Earth. Give students Worksheet 4 to complete their observations and conclusions.

(As the moon orbits the Earth, the phases change. The phase of the moon is how much of the moon appears to us on Earth to be lit up by the sun. Once a month, the moon orbits around the Earth. The phases of the moon are a new moon, waxing crescent, first quarter, waxing gibbous, full moon, waning gibbous, third quarter, and waning crescent.

As the moon begins to orbit the Earth, we can only see a portion of the lit up side. As the moon continues to orbit the Earth, we see more of the lit up side until finally the moon is on the opposite side of the Earth from the sun and we get a full moon. When we can see 100% of the lit up side, this is a full moon. As the moon continues to orbit the Earth, we see less of the lit up side. When we can't see any of the lit up side, this is called a new moon. In the new moon phase, the moon is between Earth and the sun.)

OTM2157   ISBN: 9781770783638
© On The Mark Press

3. Explain to students that they will embark on a moon watch for one month. Each night, they will observe and record what the moon looks like. Give them Worksheet 5 and a black pencil crayon to take home. Explain to students that they are to use the black pencil crayon to shade in the part of the moon that is not showing lit up. The circles on the Worksheet represent the moon, the small square boxes inside each "day square" can be used to write in the calendar numbers for the month.

*It may be beneficial to track the moon's phases each day by going to a website that offers this information. This can be an activity that is done as a class, so that the students all go home with the same information. This could be an incentive for students to check on the moon each night in order to see if the information given on the website is accurate. Track this data on a large chart, so that it can be used for reference for a later activity.

4. After 30 days, ask students to return Worksheet 5 to school. *An option at this point is to give students time to share and compare their data with each other.* Give students Worksheet 6 to complete as they analyse the data they have collected.

5. Give students Worksheet 7. With access to the internet, they will research and record explanations of solar and lunar eclipses.

(In order for an eclipse to occur, the Moon must be in direct line with the Earth (during a full moon for a lunar eclipse or during a new moon for a solar eclipse). But, the Moon's orbit around Earth is actually tipped about 5 degrees to Earth's orbit around the Sun. So this means that the Moon is usually above or below the plane of Earth's orbit. However, a few times a year a part of the Moon's shadow falls upon the surface of the Earth and an eclipse occurs.)

6. Explain to students that they will create a mobile **of the** moon phases. Referring back to the **chart** on the moon's phases created as a class, **circle** the point in the month's cycle where **a new** moon is evident, then do the same for a **waxing crescent, first quarter, waxing gibbous, full moon, waning gibbous, third quarter, and waning crescent.** Students can refer to **these** as they complete their moon phase **creation** cards for their mobiles. Give students Worksheets 8, 9, and 10. Read through the materials needed and what to do sections to ensure their understanding of the task. Then give them the materials to create their moon mobile!

DIFFERENTIATION:

Slower learners may benefit by being paired up with a stronger student to complete Worksheets 2 and 3. Also, it would be beneficial for these students to work in a small group with teacher direction to aid with idea flow and pacing in order to complete Worksheet 4.

For enrichment, faster learners could create a short poem about the moon. This poem could be attached to the centre of their moon phases mobile.

OTM2157   ISBN: 9781770783638
© On The Mark Press

The Phases of the Moon

The moon does not shine light, by itself. What we see when we look at the moon is the sun's light bouncing off the moon. Sometimes we only see part of the moon, sometimes we see the full moon, and other times we can't see the moon at all.

Have you ever wondered why the moon looks different in the sky some nights? Look at the diagram to see how the moon moves around the Earth.

The Moon as seen from Earth

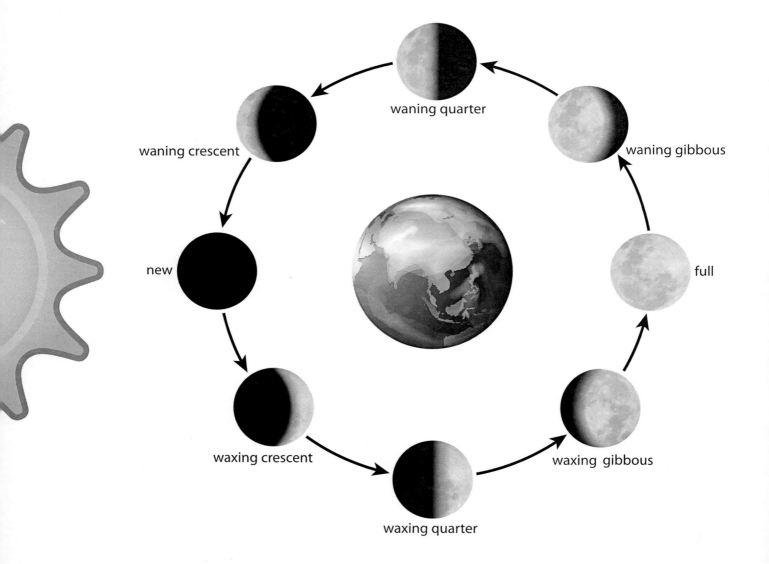

OTM2157   ISBN: 9781770783638
© On The Mark Press

Name:

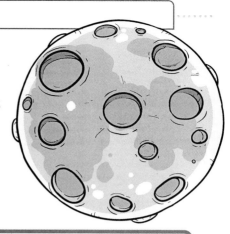

Think **Pair** **Share**

With a partner, do some thinking and sharing of ideas about the moon's phases.

Use pictures and words when recording your ideas in the chart.

Draw what the moon looks like when it is **between** the Earth and the sun.	My partner and I think it is dark because... _____ _____ _____ _____ _____ _____
Draw what the moon looks like when it is **one quarter** of the way around the Earth.	My partner and I think it is partly lit because... _____ _____ _____ _____ _____

OTM2157   ISBN: 9781770783638
© On The Mark Press

Name:

Think | Pair | Share

With your partner, continue doing some thinking and sharing of ideas about the moon's phases.

Use pictures and words when recording your ideas.

Draw what the moon looks like when the **Earth** is between it and the sun.

My partner and I think it is fully lit because...

_____

_____

_____

_____

_____

Draw what the moon looks like when it has gone **three quarters** around the Earth.

My partner and I think it is partly lit because...

_____

_____

_____

_____

_____

_____

OTM2157   ISBN: 9781770783638
© On The Mark Press

Name:

Let's Observe

When my teacher showed how the moon goes around the Earth, I noticed that…

_____

_____

_____

_____

And I noticed…

_____

_____

_____

_____

Let's Conclude

I learned that…

_____

_____

_____

_____

_____

_____

OTM2157   ISBN: 9781770783638
© On The Mark Press

Let's Go on a Moon Watch!

Observe the shape of the moon for a month. Record your observations by drawing what the moon looks like each night.

	Sunday	Monday	Tuesday	Wednesday	Thursday	Friday	Saturday
	○	○	○	○	○	○	○
	○	○	○	○	○	○	○
	○	○	○	○	○	○	○
	○	○	○	○	○	○	○
	○	○	○	○	○	○	○

OTM2157　ISBN: 9781770783638
© On The Mark Press

Name:

A Closer Look!

Let's take a closer look at the data you have **collect**ed about the moon during one month!

| **Draw** what the moon looked like when you began your observations. | **Draw** what the moon looked like after 30 days. |

Describe exactly how the moon changed during the month.

_____

_____

_____

_____

_____

_____

_____

_____

_____

Eclipsed!

Use the internet to research the answers to the following questions.

What is a lunar eclipse?

_____

_____

_____

A detailed diagram of a lunar eclipse:

What is a solar eclipse?

_____

_____

_____

A detailed diagram of a solar eclipse:

OTM2157   ISBN: 9781770783638
© On The Mark Press

Making a Mobile of the Moon

You have learned about what the phases of the moon look like. Now let's create a mobile of these moon phases!

Materials Needed:

- a coat hanger
- scissors
- some yarn
- a ruler
- moon phase name cards
- a black pencil crayon
- moon phase creation cards (2 sheets)

What to do:

1. Cut out the moon phase name cards and creation cards.

2. Look at the moon phase chart that your class completed. Find the 8 different phases of the moon that your teacher circled.

3. Draw one phase on each moon phase creation card. (Use a black pencil crayon to shade in the part of the moon that is not lit up).

4. Punch a hole at the top of each of the moon phase creation cards.

5. Punch a hole at the top and at the bottom of each of the moon phase name cards.

6. Cut 16 pieces of yarn, each about 8 inches (20 cm) long.

7. Tie a piece of yarn to the top of each moon phase creation card. Then tie the other end of the pieces of yarn to the bottom of the matching moon name cards.

8. Tie a piece of yarn to the top of each of the moon name cards. Then tie the other end of the pieces of yarn to the coat hanger. You have created a moon mobile!

OTM2157 ISBN: 9781770783638
© On The Mark Press

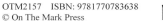

Moon Phase Name Cards

New Moon	**Waxing Crescent**
First Quarter	**Waxing Gibbous**
Full Moon	**Waning Gibbous**
Third Quarter	**Waning Crescent**

OTM2157 ISBN: 9781770783638

Name:

Moon Phase Creation Cards

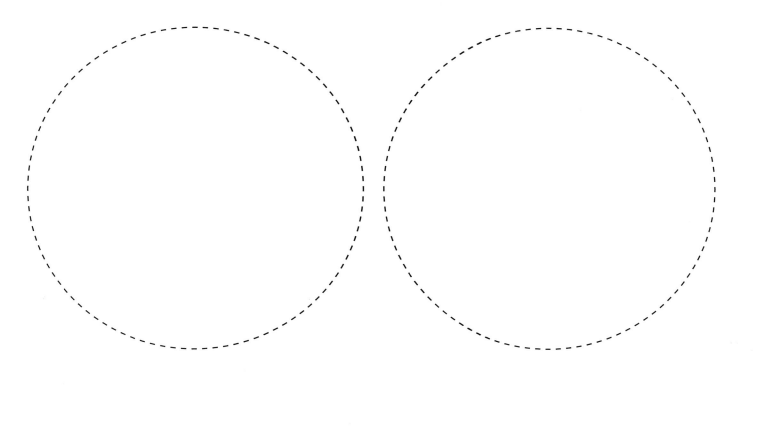

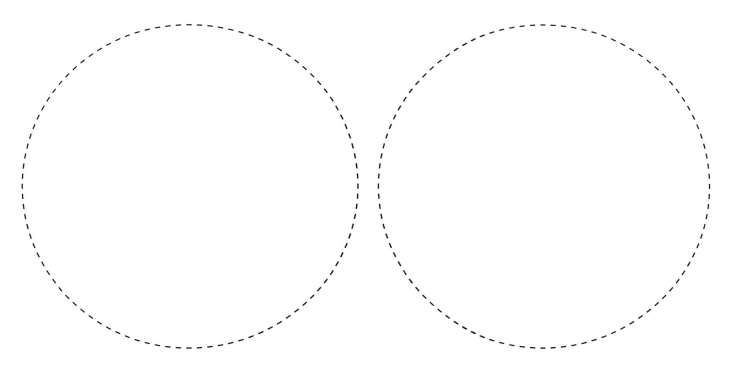

OTM2157   ISBN: 9781770783638

CONSTELLATIONS

LEARNING INTENTION:

Students will learn about the constellations in our night sky.

SUCCESS CRITERIA:

- identify and label constellations in our night sky
- research a constellation to determine its position, special features, and story behind it
- use a Venn diagram to compare and contrast the Greek myth and Aboriginal legend of the Big Dipper's existence in the sky
- create a legend for a constellation, illustrate it in graphic text form

MATERIALS NEEDED:

- a copy of "Stars in the Night Sky" Worksheet 1 for each student
- a copy of "Constellations in the Night Sky" Worksheet 2 and 3 for each student
- a copy of "A Stargazing Report!" Worksheet 4 and 5 for each student
- a copy of "How the Big Dipper Came to Be" Worksheet 6 for each student
- a copy of "A Legend in the Making!" Worksheet 7 and 8 for each student
- a lamp, a long extension cord
- access to computers with internet connection
- overhead projector or document camera (optional)
- clipboards, markers, pencil crayons, pencils

PROCEDURE:

*This lesson can be done as one long lesson, or be divided into four shorter lessons.

1. Using Worksheet 1, do a shared reading activity with the students. This will allow for reading practice and learning how to break down word parts in order to read the larger words in the text. Along with the content, discussion of certain vocabulary words would be of benefit for students to fully understand the passage.

Some interesting vocabulary words to focus on are:

- galaxy
- particles
- hydrogen
- helium
- constellation
- dimmest
- kelvin
- surface
- extremely
- life span

2. Give students Worksheet 2 (or project it on a large screen using an overhead projector or document camera). Discuss the different constellations and the different positions they take in the night sky depending on which month of the year it is. To help students understand why the constellations change position, try the following exercise:

- set a lamp in the centre of an empty space (the lamp will represent the sun)
- have students stand in a large circle around the lamp, facing away from it
- tell students that their shadow is the night sky and ask them to imagine that the walls are covered in stars
- tell students to slowly walk in a circle around the lamp; the spot on the wall that their shadow is covering is changing as they move, just as the Earth's night sky changes as it orbits the sun

(Students may be interested to know that the stars we see in the sky depend on where in the world we are. The stars we see in the Northern Hemisphere in a certain month of the year are different from the ones that can be seen in the Southern Hemisphere in that same month of the year. We know that the seasons in the year reverse between the hemispheres; this is the same case for the stars we are able to see in the sky.)

OTM2157   ISBN: 9781770783638
© On The Mark Press

3. Give students Worksheet 3 to complete. Encourage students to do some "star gazing" on their own one night. A suggestion is to have them draw a few that they recognize, then bring the drawings in to share with the class.

4. Give students Worksheets 4 and 5, along with a clipboard and pencil. For this activity, students will need computers and access to the internet in order to research a constellation of their choice.

5. Explain to students that many of the constellations have myths *and* legends to explain how they came to be in the sky. As an example, use Ursa Major (the Big Dipper). You can access the internet to find a Greek myth and an Aboriginal legend to share with the students. As an alternative, you can purchase National Geographic's My First Pocket Guide Constellations (for Greek myth), and C.J. Taylor's All the Stars in the Sky (for Aboriginal legend). Students will compare and contrast the two explanations by completing the Venn diagram on Worksheet 6. Follow up with a large group discussion if time permits.

6. Give students Worksheets 7 and 8. They will use the constellation that they researched in an earlier activity to create a legend of their own that explains how the constellation came to be in our night sky.

DIFFERENTIATION:

Slower learners may benefit by having a pre-determined "short list of constellations" to choose from in order to complete the research activity on Worksheets 4 and 5. Pairing up with a peer to complete this activity is another option. A further accommodation is to eliminate Worksheet 7 for these students and have them create a graphic text on Worksheet 8 of the Aboriginal legend of the Big Dipper that was shared with the entire class for the Venn diagram activity.

For enrichment, faster learners could recreate their legend using computer software called Comic Life. This is available for Mac and PC systems.

OTM2157   ISBN: 9781770783638
© On The Mark Press

Name:

Stars in the Night Sky

Did you know that our sun is actually a star? It is just one of the many millions of stars in our galaxy!

A star is a glowing ball of dust and gases like hydrogen and helium that continues to grow as it pulls in more particles in space. The core of a star gets extremely hot because of the pressure. Some stars are brighter and hotter than others, and they come in different colors.

Stars can be blue, white, yellow, or red. The hottest and brightest stars are blue, then white ones, yellow ones are slightly cooler, and the coolest and dimmest are red. Because red stars are the coolest, they burn very slowly and have a longer life span than blue or white stars.

Fast Fact!

A group of stars that appear to form a picture in the night sky is called a constellation. There are 88 recognized constellations that cover the entire sky.

The more mass a star has, the hotter its temperature. A star that is big, bright, and blue is extremely hot. It is called a blue giant.

A blue giant can range in temperature from 20 000 to 50 000 kelvin. To compare this, a yellow star like the sun has a surface temperature of about 6000 kelvin. So a blue giant star is blazing hot!

OTM2157   ISBN: 9781770783638

Constellations in the Night Sky

There are many constellations in the night sky. We use constellations to determine an area in the celestial sky. Different constellations are visible at different times of the year. They change in position in the sky depending on what time of year it is.

Below are diagrams of what a clear night sky in the Northern Hemisphere looks like during the year.

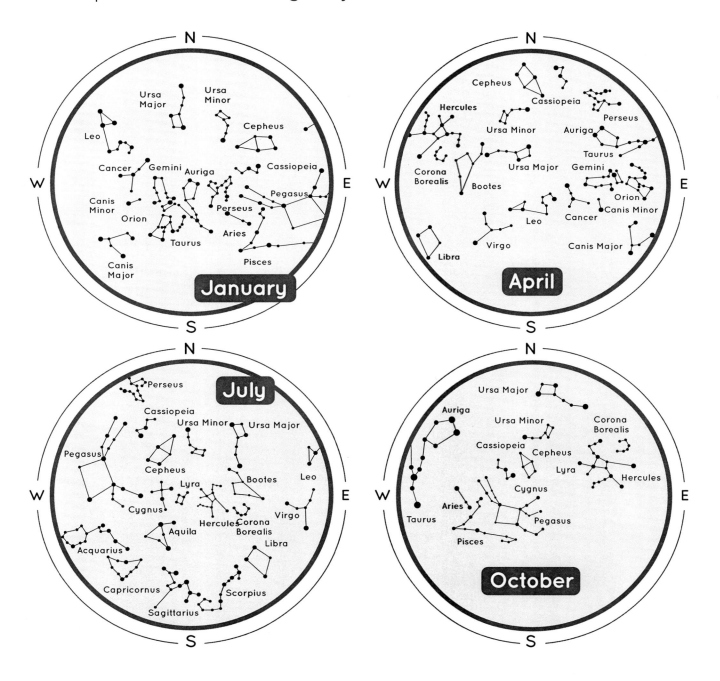

Name:

It is time to draw some constellations that are in our night sky! Use the diagrams on Worksheet 2 and the clues below to help you.

Ursa Major (the Big Dipper).
Clue: On a winter night, I sit high in the northern sky.

Orion
Clue: In the early spring time, you will find me in the western part of the sky.

Ursa Minor (the Little Dipper).
Clue: On a summer night, I am between Ursa Major and Cassiopeia.

Cassiopeia
Clue: On a clear autumn night, you will find me beneath the "Little Dipper".

OTM2157   ISBN: 9781770783638
© On The Mark Press

Name: _____

A Stargazing Report!

You have learned that there are 88 recognized constellations in our night sky. Now it is time to choose one to study and write a stargazing report!

Using the internet, research a constellation of your choice. Here is a graphic organizer to guide you through this task.

This is an illustration of the constellation: _____

Explain when your constellation can be seen in the night sky, including from which hemisphere.

_____

_____

_____

_____

_____

OTM2157   ISBN: 9781770783638
© On The Mark Press

Name: _____

Tell about the special features of your constellation. For example, does it contain any special stars?

_____

_____

_____

_____

Name a few constellations that are near the constellation that you chose.

_____

_____

_____

_____

Each constellation has its own **myth**. What is the story behind your constellation?

_____

_____

_____

_____

_____

_____

_____

_____

_____

OTM2157   ISBN: 9781770783638
© On The Mark Press

Name:

How the Big Dipper Came to Be

Greek mythology is only one source to help **us** learn about the story behind a cluster of stars in the sky. There are **also** Aboriginal legends written about how the constellations appeared **in** the sky.

Use the Venn diagram below to compare and **contrast** a Greek myth and an Aboriginal legend about how the Big Dipper **came** to be in our night sky.

Remember the middle of the Venn diagram is used to list similarities.

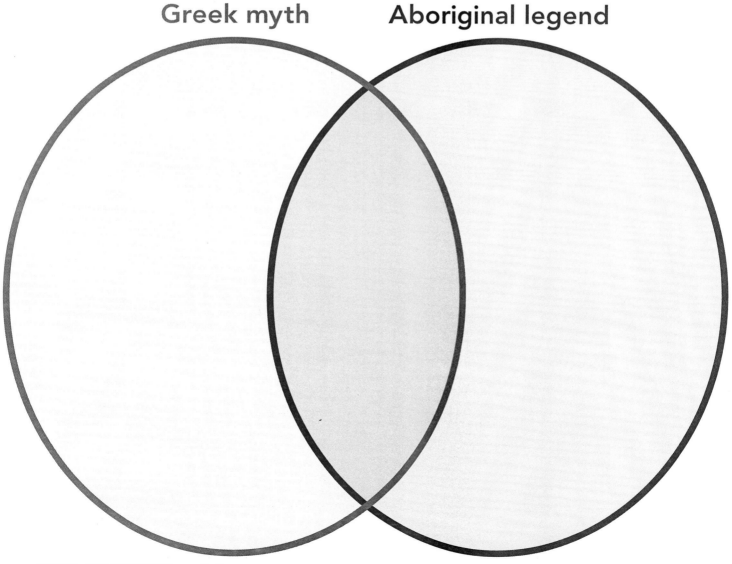

Greek myth   Aboriginal legend

OTM2157   ISBN: 9781770783638
© On The Mark Press

A Legend in the Making!

Use the constellation that you researched on Worksheets 4 and 5 to create a **legend of your own** to explain how it came to be in our night sky.

Use the organizer below to help you with your thinking. Then, create your legend in graphic text form on Worksheet 8.

Title: _____

Characters: _____

_____

_____

The Problem: _____

_____

_____

Sequence of Events:

1. _____

2. _____

3. _____

4. _____

5. _____

6. _____

The Solution: _____

_____

_____

_____

_____

_____

OTM2157   ISBN: 9781770783638

Name:

ASTEROIDS, COMETS, & METEOROIDS

LEARNING INTENTION:
Students will learn about the unique features of asteroids, comets, and meteoroids.

SUCCESS CRITERIA:
- research, record, and describe facts about asteroids
- research, record, and describe facts about comets
- illustrate the transformation of a meteoroid as it enters a planet's atmosphere
- research and record facts about features in our solar system through written descriptions and illustrations

MATERIALS NEEDED:
- a copy of "Inside the Asteroid Belt" Worksheet 1 for each student
- a copy of "Tailing the Comets!" Worksheet 2 for each student
- a copy of "From Meteoroids to Meteorites!" Worksheet 3 for each pair of students
- a copy of "An Orbiting You Go!" Worksheet 4 and 5 for each student
- access to the internet
- markers, chart paper, pencils, pencil crayons

PROCEDURE:
This lesson can be done as a long lesson, or divided into three or four shorter lessons.
1. Give students Worksheet 1. With access to the internet, students will research facts about asteroids. Navigating students to www.nasa.gov to explore the NASA website would be beneficial for their search. Once students have acquired five facts, they can pair up with a classmate to share their findings. An option is to come together as a large group to continue the discussion and possibly record facts on chart paper that could be posted in the classroom for future reference.

2. Give students Worksheet 2. With access to the internet, students will research facts about comets. Navigating students to www. nasa.gov to explore the NASA website would be beneficial for their search. Once students have acquired five facts, they can pair up with a classmate to share their findings. An option is to come together as a large group to continue the discussion and possibly record facts on chart paper that could be posted in the classroom for future reference.

3. Give students Worksheet 3. They will read through the description of how a meteoroid becomes a meteorite, then use this information to illustrate the process of this change. Students will also formulate a question that they would like to research. Once the research is completed, provide students an opportunity to share their questions and answers with the large group. The facts about this topic that get uncovered could be recorded on chart paper and posted in the classroom for future reference.

4. Give students Worksheets 4 and 5. They will use the internet to research information about asteroids, comets, and meteoroids in our solar system.

DIFFERENTIATION:
Slower learners may benefit by researching and recording only 2 or 3 facts for activities on Worksheets 1 and 2. An additional accommodation is to have these learners work with a strong peer in order to 'navigate' the research required to complete Worksheets 4 and 5.

For enrichment, faster learners could work in a small group to create a skit called, "A Meteorite Landed in My Schoolyard!" or "My Ride on a Comet's Tail". These titles could also be used for a story writing assignment.

OTM2157   ISBN: 9781770783638

Inside the Asteroid Belt

An asteroid is a solid, rocky, cratered mass that revolves around the sun. Most asteroids can be found orbiting in a part of space known as the Asteroid Belt. The Asteroid Belt is located between the orbits of Mars and Jupiter.

Get the Facts!

Use the internet to research five fast facts about asteroids. Record your findings below.

1. _____

2. _____

3. _____

4. _____

5. _____

Discuss with a classmate the facts that you discovered about asteroids.

Tailing the Comets!

A comet is like a snowball of frozen gases, rock, and dust. As a comet orbits the sun, its proximity to the sun allows the sunlight to force some gases and dust out of the head of the comet, causing a bright tail to appear behind it.

Get the Facts!

Use the internet to research five fast facts about comets. Record your findings below.

1. _____

2. _____

3. _____

4. _____

5. _____

Discuss with a classmate the facts that you discovered about comets.

Did you know?

Comets travel in orbits like an elongated egg shape, making loops around the sun. Scientists can calculate the path a comet will travel in, and its distance. This is how they are able to figure out when a comet will pass by our planet Earth!

OTM2157   ISBN: 9781770783638
© On The Mark Press

From Meteoroids to Meteorites!

From meteoroid to meteorite, let's track this development!

Meteoroids are particles of rock or metal in the solar system.

▼

When a meteoroid comes in contact with a planet's atmosphere, it becomes a meteor which looks like a shooting star in the night sky.

▼

The pieces of the meteor that fall to the surface of a planet are called meteorites. A meteorite looks like burned iron rock with big holes in it.

Can you illustrate this transformation?

Formulate a question about meteoroids, meteors, or meteorites that you would like to research.

Name:

An Orbiting You Go!

You are on a space mission to find out more about what is orbiting in our space. The internet is a galaxy of information. Away you go!

On a Mission!

What is Ceres and where is it?

_____

_____

_____

The next phase of your mission:

Find out about comet Halley's next visit past Earth.

_____

_____

_____

_____

What can you discover about the Kuiper Belt?

_____   ┌─────────────────────────────┐
                                           │ An illustration of the Kuiper Belt: │
_____   │   │
                                           │   │
_____   │   │
                                           │   │
_____   │   │
                                           │   │
_____   │   │
                                           │   │
_____   │   │
                                           │   │
_____   └─────────────────────────────┘

OTM2157   ISBN: 9781770783638
© On The Mark Press

What can you discover about the Oort Cloud?

_____

_____

_____

_____

_____

_____

_____

An **illustration** of the Oort Cloud:

The final phase of your mission:

Give a detailed explanation of the Leonids.

_____

_____

_____

_____

_____

When to watch for them on planet Earth: _____

Challenge question:

What is a satellite and what purpose could it have?

_____

_____

_____

_____

OTM2157   ISBN: 9781770783638
© On The Mark Press

SPACE EXPLORATION

LEARNING INTENTION:

Students will learn about the technology needed for space exploration, and the costs and benefits of this science.

SUCCESS CRITERIA:

- describe different types of technology used for space exploration
- research, record, and discuss details of a space exploration mission
- listen to and record facts about space exploration
- research, discuss, and evaluate the costs vs. the benefits of space exploration

MATERIALS NEEDED:

- a copy of "Space Technologies" Worksheet 1, 2, and 3 for each student
- a copy of "A Detailed Mission" Worksheet 4 and 5 for each student
- a copy of "Costs vs. Benefits" Worksheet 6 for each pair of students
- access to the internet
- markers, chart paper
- clipboards, pencils, pencil crayons
- egg cartons, Styrofoam trays or plates, boxes, plastic bottles, milk containers, newspaper, aluminum foil, cardboard tubes, margarine tubs or other plastic containers, glue, string, fasteners, paint, paint brushes, scissors *(optional materials)*

PROCEDURE:

***This lesson can be done as a long lesson, or divided into four shorter lessons.**

1. Using Worksheet 1, 2, and 3, do a shared reading activity with the students. This will allow for reading practice and learning how to break down word parts in order to read the larger words in the text. Along with the content, discussion of certain vocabulary would be of benefit for students to fully understand the passage.

   Some interesting vocabulary words to focus on are:

   - launch
   - manned
   - astronomer
   - meteorology
   - mission
   - unmanned
   - laboratory
   - transportation system
   - retired
   - versatile
   - solar arrays
   - technological advances

2. Give students Worksheet 4 to complete. With access to the internet, students will research a space exploration mission. Navigating students to www.nasa.gov to explore the NASA website would be beneficial for their research.

3. Divide students into small groups of 3 or 4. They will share the details of the mission that they researched with their group members. Group members will record interesting facts that they learn about throughout the discussion, on Worksheet 5. An option is to come together as a large group to continue the discussion and possibly record facts on chart paper that could be posted in the classroom for future reference. Group members can then engage in the suggested discussion item at the bottom of Worksheet 5.

OTM2157   ISBN: 9781770783638
© On The Mark Press

4. Engage students in a discussion about the costs and the benefits of space exploration to society and to the environment. Record ideas on chart paper for later reference. Give students Worksheet 6. Explain to them that they will evaluate the costs vs. the benefits of space exploration. To do this, they can:

- consider the ideas that were discussed in the large group

- use the internet to research costs and benefits of space exploration

- refer back to information that they learned in the small group sharing activity in item #3

DIFFERENTIATION:

Slower learners may benefit by working as a small group with teacher support to participate in the group talk activity on Worksheet 5. This would allow for only the pertinent information to be shared, discussed, and recorded. Facts learned from this activity could be recorded on one piece of chart paper, which would eliminate the need to have these learners fill out Worksheet 5 on their own.

For enrichment, faster learners could create a diagram of a spacecraft or space instrument. Then they could build it using recyclable materials and classroom consumables.

Name:

Space Technologies

There are different technologies that are being used for space exploration to date. Some of these include telescopes, probes, space shuttles, and a space station. Let's read about some of the more popular technological advances that have been made in the area of space exploration!

The Hubble Space Telescope is one of the largest and most versatile telescopes that have been launched into space. Named after the astronomer Edwin Hubble, the Hubble Space Telescope was built by NASA with input from the European Space Agency. It was launched in 1990 and continues to be in operation.

The Hubble is in orbit around our Earth. It captures high resolution images of objects in space, which have provided scientific data to be used in space study. The Hubble is the only telescope that has ever been designed to be serviced in space. Astronauts have gone on missions over the years to replace instruments on the Hubble telescope in order for it to maintain its effectiveness in space exploration.

A probe is a robotic spacecraft that, once launched, leaves the Earth's orbit to explore space objects and other planetary bodies. There are no astronauts aboard these types of spacecrafts.

The Voyager 1 Space Probe near Jupiter.

The Voyager 1 space probe launched in 1977, and it continues to be in operation. It has visited Jupiter and Saturn, and provided detailed pictures of their moons for scientists to study. Of all the man-made space objects, the Voyager 1 is currently travelling the farthest from the Earth.

OTM2157   ISBN: 9781770783638
© On The Mark Press

The Space Shuttle was a space transportation system that was operated by NASA from 1981 to 2011, at the Kennedy Space Center in Florida, USA. The Space Shuttle had been used to launch 135 missions into space.

Those missions included satellites, probes, the Hubble Space Telescope, and some space shuttles. The space shuttles were built to take astronauts into the Lower Earth Orbit, which is between 160km – 2000km above the surface of the Earth.

The first shuttle that the Space Shuttle launched was the Enterprise, which was only built for approach and landing tests, and did not fly into orbit. This was followed by the launch of more fully operational space shuttles, these being the Columbia, the Challenger, the Discovery, the Atlantis, and the Endeavor.

NASA's Space Shuttle Endeavor at the Kennedy Space Center launch pad, preparing for a mission to the International Space Station.

© Gary Blakeley | Shutterstock.com

Unfortunately, the Challenger and the Columbia were destroyed before they were able to reach their missions in 1986 and in 2003. At the end of the space shuttle Atlantis' flight in July 2011, the Space Shuttle was retired from service.

The Space Shuttle will be replaced by the **Orion Multi-Purpose Crew Vehicle**. The Orion spacecraft will be able to take crews of astronauts beyond the Lower Earth Orbit and into deep space to such places as the Moon, near Earth asteroids, and Mars!

The Orion spacecraft is scheduled to make an unmanned test flight in 2014, with the first manned mission expected to take place after 2020.

OTM2157   ISBN: 9781770783638
© On The Mark Press

The International Space Station is a space research laboratory that is located in the Lower Earth Orbit about 400km above the Earth's surface. The ISS is traveling in orbit at about a speed of 7 km per second.

The International Space Station orbits the Earth about 16 times each day.

The International Space Station is the largest spacecraft in orbit. This modular structure is about as long as a football field. Power for the ISS is supplied by the many solar arrays aboard its structure.

The International Space Station Program brings together flight crews from space agencies from USA, Russia, Europe, Japan, and Canada. Since the year of 2000, crews of astronauts have lived and worked inside the space station. The crew members not only maintain the operation of the space station, they also conduct experiments in many scientific fields such as biology, physics, meteorology, and astronomy.

Fast Fact!

The average stay for an astronaut at the International Space Station is about 6 months. In 2015, for the first time, a couple of astronauts will stay at the ISS for one year. This is so scientists can learn how the human body reacts to a long period of exposure to microgravity. This study is being conducted as preparations are being made to send astronauts out into the solar system on the Orion Multi-Purpose Crew Vehicle.

OTM2157   ISBN: 9781770783638
© On The Mark Press

Name:

A Detailed Mission

You have learned that different types of spacecrafts and space instruments have gone on explorations of our solar system. Your task now is to research and provide details of a space exploration mission.

Let's Research!

Name of spacecraft or instrument: _____

Date of launch: _____

Launch site: _____

Mission destination: _____

Purpose of mission:

- _____
- _____
- _____
- _____

Date of return: _____

Diagram of spacecraft/instrument:

OTM2157   ISBN: 9781770783638
© On The Mark Press

Group Talk!

Listen to your group members as they share details of the space exploration mission that they researched. Record interesting facts that you learn about.

- _____
  _____
  _____

- _____
  _____
  _____

- _____
  _____
  _____

- _____
  _____
  _____

- _____
  _____
  _____

- _____
  _____
  _____

- _____
  _____

Share your thoughts with your group members about where you think the next mission in our solar system should be, and explain your thinking.

OTM2157 ISBN: 9781770783638
© On The Mark Press

Name:

Costs vs. Benefits

Taking into account different points of view, evaluate the costs and the benefits of space exploration. Organize your thinking in the t-chart below.

Costs	Benefits

In your opinion, are the benefits of space exploration worth the costs? Explain your thinking.

_____

_____

_____

_____

_____

OTM2157   ISBN: 9781770783638
© On The Mark Press

SPACE EXPLORERS

LEARNING INTENTION:
Students will learn about learn the contributions to space exploration, and about the technological advances that allow humans to adapt to life in space.

SUCCESS CRITERIA:
- research and record facts about the contributions of a space exploration pioneer
- research and record features of the International Space Station
- research and record facts about human adaptation to living life in space

MATERIALS NEEDED:
- a copy of "Contributing to Space Exploration" Worksheet 1 and 2 for each student
- a copy of "All in a Day's Work!" Worksheet 3, 4, 5, 6, and 7 for each student
- access to the internet
- clipboards, pencils, pencil crayons

PROCEDURE:
***This lesson can be done as a long lesson, or divided into shorter researching periods.**
1. Give students Worksheets 1 and 2. With access to the internet, students will research a person who has contributed in some way to space exploration. Contributions could be in the form of involvement in a project that has enhanced the advances of space exploration, or it could be about the participation in a space exploration mission. The finished assignments could be displayed on a bulletin board to highlight the students' learning of the solar system and space exploration.

2. Give students Worksheets 3, 4, 5, 6, and 7. With access to the internet, students will research the type of work astronauts do in space and how they manage their personal daily functions while aboard a spacecraft for an extended period of time. Navigating students to **www.asc-csa.gc.ca/eng** to explore the Canadian Space Agency website or to **www.nasa.gov** to explore the NASA website would be beneficial for their research.

Teaching options at this point:
- using an iPod, iPad, Macbook, or video camera, set up an oral response recording station in the classroom, student could take turns recording their oral responses to the 'consider this' section on Worksheet 7. The video clips could be viewed later in a large group setting, with follow up discussion/debate.
- a follow-up option to the written component of Worksheet 7 is to have students discuss as a large group their findings that they have completed. This would allow the opportunity for some rich discussion and debating to occur.

DIFFERENTIATION:
Slower learners may benefit by working with a strong peer in order to 'navigate' the research required to complete Worksheets 3, 4, 5, and 6. Sections of the report could be divided between the partners to lessen the work load. An additional accommodation for these learners is to come together in a small group with teacher direction to orally complete Worksheet 7, thus eliminating the written expectation.

For enrichment, faster learners could continue their research by investigating the different work roles and locations throughout the world that are needed to support human spaceflight programs.

OTM2157   ISBN: 9781770783638
© On The Mark Press

Contributing to Space Exploration

There have been a lot of technological advances in space exploration, due to the interest and commitment of certain people.

Your task now is to research a person who has contributed to space exploration. Ideas for this task could be an astronaut, a scientist, or an engineer. Use the organizer below as a guide for your research.

Let's Research!

Name of person: _____

Occupation: _____

Launch site: _____

Name of Mission/Project involvement:

_____

_____

Purpose of mission or project:

- _____
- _____
- _____
- _____

Outcome of mission or project: _____

_____

_____

_____

_____

_____

_____

OTM2157   ISBN: 9781770783638
© On The Mark Press

Illustration of spacecraft/instrument involved:

In **your** opinion, how has this mission or project provided more information about space or helped to advance space exploration?

_____

_____

_____

_____

_____

If you had the opportunity to ask the person you chose to research two questions, what would they be? Record them below.

1) _____

_____

2) _____

_____

OTM2157　ISBN: 9781770783638
© On The Mark Press

All in a Day's Work!

Have you wondered about the work astronauts **do** during a stay at the International Space Station? Have you wondered how they manage the necessities of daily living while they are floating in space?

Your task now is to research what daily life is **like** for the astronauts at the International Space Station. Use the organizer **below** as a guide for your research.

Let's Research!

Using an illustration **and** a written description, **describe** the inside of the International Space Station.

Inside the Space Station

_____

_____

_____

_____

_____

OTM2157   ISBN: 9781770783638
© On The Mark Press

Astronauts are involved in many projects during their stay at the International Space Station. Explain in detail, one of these projects.

_____

_____

_____

_____

_____

Give an example of a situation in which an astronaut would have to venture outside of the Space Station.

_____

_____

_____

_____

_____

_____

_____

_____

_____

_____

_____

_____

_____

_____

_____

_____

Diagram of a spacesuit worn by an astronaut:

OTM2157   ISBN: 9781770783638
© On The Mark Press

Name:

What is on the daily menu for an astronaut to eat?

Breakfast

_____
_____
_____
_____

Lunch

_____
_____
_____
_____

Dinner

_____
_____
_____
_____

Snacks

_____
_____
_____
_____

Explain how the food at the Space Station is stored and prepared.

Food Storage:

_____
_____
_____
_____

Food Preparation:

_____
_____
_____
_____
_____

OTM2157   ISBN: 9781770783638

An illustration of an astronaut's sleeping quarters aboard the Space Station:

Explain why exercise is important for an astronaut living aboard the Space Station.

_____

_____

_____

_____

_____

_____

Describe how an astronaut brushes his/her teeth aboard the Space Station.

_____

_____

_____

_____

_____

_____

OTM2157   ISBN: 9781770783638
© On The Mark Press

Consider This!

Life in space can be challenging for humans. **Certain** technologies exist and are being used to overcome the **challenges** of living in space.

Make a list of the technologies that the astronauts living aboard the Space Station are using to help them mimic **living** conditions on Earth.

- _____
- _____
- _____
- _____
- _____
- _____
- _____
- _____

Could robots replace the role of humans aboard spacecrafts in space? Explain your thinking.

_____
_____
_____
_____
_____

If you had the opportunity to ask an astronaut aboard the International Space Station one question, what would it be?

_____
_____
_____
_____
_____

OTM2157   ISBN: 9781770783638
© On The Mark Press

GOING TO THE EXTREME!

LEARNING INTENTION:

Students will learn about the exploration of different extreme environments.

SUCCESS CRITERIA:

- brainstorm and record examples of extreme environments
- complete a KWL chart about an extreme environment
- research and record facts about the exploration of an extreme environment
- organize and display research in a visual display
- conduct an oral presentation about an extreme environment

MATERIALS NEEDED:

- a copy of "Extreme Environments" Worksheet 1 for each student
- a copy of "Know, Wonder, and Learn!" Worksheet 2 for each student
- a copy of "Adventuring into the Extreme!" Worksheet 3, 4, 5, and 6 for each student
- access to the internet
- Bristol board (a piece for each student)
- glue, scissors, clipboards, chart paper, markers
- pencils, pencil crayons

PROCEDURE:

*This lesson can be done as a long lesson, or divided into shorter researching periods.

1. Discuss the meaning of "extreme" and "extreme environment" with students to ensure their understanding of these terms. Divide students into pairs and give each Worksheet 1. Students will work together to brainstorm some types of extreme environments. After sufficient time, come back together as a large group to share ideas. Record responses on chart paper, then post it in the classroom for future reference.

Extreme environments are places people don't live in, but choose to explore. For example:

- polar regions
- space
- depths of the ocean
- volcanoes
- deserts
- mountain ranges
- rain forests
- African savannas
- the atmosphere (flight)
- ocean tide pools
- caves

2. Give students Worksheet 2. They will choose one of the extreme environments that they listed on Worksheet 1. Students will record what they already know about it, and what they wonder about it. (The 'What I Have Learned' section will be completed after students have the opportunity to research the environment that they chose.)

3. Give students Worksheets 3, 4, 5, and 6. With access to the internet, students will research the extreme environment that they chose on Worksheet 2. Along with providing a type and location of an extreme environment, students will research how it is accessed, what can be studied there, exploration tools used, dangers in the environment, protective gear used, discoveries that have been made, and potential discoveries to occur.

4. Students can complete the 'What I Have Learned' section on Worksheet 2.

5. Students can display their finished work on Bristol board. Then they can practice an oral presentation format of the information they have displayed. Divide students into small groups so that they can present their extreme environment to their peers. This would allow for students to learn about other extreme environments.

OTM2157   ISBN: 9781770783638
© On The Mark Press

DIFFERENTIATION:

Slower learners may benefit by working as a small group with teacher support and direction to complete Worksheets 2, 3, 4, 5, and 6. Worksheet 2 could be done together on chart paper.
Then sections of the research for the extreme environment could be divided between these students to lessen the work load. The final product could be displayed on one Bristol board, in which these students would then orally present their section to the large group.

For enrichment, faster learners could create a slideshow presentation of their research and present it to the members of their small group in that format. *An alternative accommodation is to have these learners research the type of occupations that may be involved in the exploration process of the extreme environment that they chose to research.*

OTM2157   ISBN: 9781770783638
© On The Mark Press

Name:

Extreme Environments

Think **Pair** **Share**

With a partner do some thinking and sharing of examples of extreme environments on our planet Earth. In the box below, record your ideas.

OTM2157   ISBN: 9781770783638
© On The Mark Press

Know, Wonder, and Learn!

Choose one of the extreme environments that you listed on Worksheet 1. In the chart below, record some things that you already know about that environment, and what you wonder about that environment. After some study of this extreme environment, you will be able to record what you have learned about it.

What I Already Know	What I Wonder	What I Have Learned

OTM2157   ISBN: 9781770783638
© On The Mark Press

Name:

Adventuring into the Extreme!

You have already chosen an extreme environment. Now it is time do some research in order to, learn more about it. Use the organizer below as a guide for your research. Let's start exploring!

Let's Research!

Extreme Environment: _____

Location: _____

A detailed illustration of the environment:

OTM2157 ISBN: 9781770783638
© On The Mark Press

Describe how humans access this environment.

_____

_____

_____

_____

_____

_____

Explain what can be studied and learned in this environment.

_____

_____

_____

_____

_____

_____

_____

Tools/Instruments used to explore this environment:

- _____
- _____
- _____
- _____
- _____
- _____
- _____
- _____
- _____

Explain the possible dangers that people face while in this environment.

_____

_____

_____

_____

_____

_____

Gear people use to protect themselves while in this environment:

- _____
- _____
- _____
- _____
- _____
- _____
- _____

Diagram of a piece of protective gear:

OTM2157   ISBN: 9781770783638
© On The Mark Press

List some of the discoveries that have been **made** about this environment.

- _____
  _____

- _____
  _____

- _____
  _____

- _____
  _____

- _____
  _____

- _____
  _____

According to scientists and other experts, what is still to be discovered about this extreme environment?

_____

_____

_____

_____

_____

_____

_____

_____

OTM2157   ISBN: 9781770783638
© On The Mark Press

TECHNOLOGY IN THE EXTREME

LEARNING INTENTION:

Students will learn about learn about technologies used for extreme environment exploration.

SUCCESS CRITERIA:

- embark on a nine day expedition to explore the planet Mars
- observe and record features of Mars using available technology
- research and record technological advancements of a tool used in an extreme location
- research a technological advancement made by your country as a contribution to the exploration of an extreme environment
- debate the effects of responsible uses of technology vs. the misuses of technology

MATERIALS NEEDED:

- a copy of "An Exploration Expedition!" worksheet 1, 2, 3, and 4 for each student
- a copy of "A Time Line of Technology" worksheet 5, 6, and 7 for each student
- a copy of "A Contribution from Home" worksheet 8 for each student
- binoculars, a telescope
- a camera, an iPod, or an iPad
- a paper mache balloon to represent the planet Mars, string or rope
- access to the internet
- clipboards, pencils, pencil crayons, chart paper, markers
- modeling clay, sculpting tools *(optional materials)*

PROCEDURE:

Item 1 in this lesson should span over nine days. Items 2, 3, and 4 can be done as one a long lesson, or be divided into three or four shorter lessons.

1. The purpose of the following exercise is to demonstrate for students that with technological advances, exploration of extreme environments has become easier and has led to further discoveries being made about them.

Cover a balloon with paper mache, and ensure that its surface is rough. Paint it a rusty red color, with some dark patchy areas, and white at its polar regions. This will be a representation of the planet Mars. If it is possible, suspend this object somewhere in the school yard, so that there is a great distance between it and where the students will stand to begin their exploration of this extreme environment, which will be done in stages (over 9 days).

Give students a pencil, clipboard, and Worksheets 1, 2, 3, and 4. They will go outside to begin their exploration expedition of Mars.

Stage 1 (day one)
The students can only look at the object from a far distance

Stage 2 (day two)
The students can look at the object from the same distance as on day one, but can use the technology of binoculars

Stage 3 (day three)
The students can look at the object from the same distance as on day one, but can use the technology of a telescope

Stage 4 (day four)
One student will act as a space probe that has entered the atmosphere of Mars, by travelling half way from the starting point that they have

OTM2157 ISBN: 9781770783638
© On The Mark Press

had in the previous stages, towards the object, then will return to the original starting point to share with the class

Stage 5 (day five)
One student will act as a space probe that flies by Mars, by running by the object without stopping, then returning to the original starting point to share with the class

Stage 6 (day six)
One student will act as a space probe that orbits Mars, by circling the object 3 times without stopping, then returning to the original starting point to share with the class

Stage 7 (day seven)
One student will act as a space probe that orbits Mars with the ability to take pictures using a camera, iPod, or iPad; the student will circle the object 3 times without stopping, taking pictures, then return to the original starting point to share with the class

Stage 8 (day eight)
All students can visit Mars at one angle only

Stage 9 (day nine)
All students can visit Mars, exploring all areas of this environment

Upon completion of the exercise, discuss with students what technologies were used to explore the extreme environment, and discuss the impacts those technologies made in the discoveries of this environment.

2. Give students Worksheets 5, 6, and 7. With access to the internet, students will research a tool/ instrument that they had listed in the previous lesson, titled "Adventuring into the Extreme!", Worksheets 3 – 6. They will create a timeline of its major advancements, which have led to further discoveries in the extreme environment in which it is used today. Finished timelines and diagrams could be displayed on a bulletin board.

3. Give students Worksheet 8. With access to the internet, students will research a technological advancement made by their country as a contribution to the exploration of an extreme environment.

4. Engage students in a discussion/ debate about the responsible uses of technology that have led to important discoveries and scientific breakthroughs vs. the possible misuses of technology that could be having a negative effect on society and our environment.

DIFFERENTIATION:
Slower learners may benefit by only researching two or three technological advancements for the tool/instrument of their choosing on Worksheet 6. Then, they could only complete a diagram of it in its modern form on Worksheet 7. *An additional accommodation would be to work together as a small group to research their country's contribution to technology used in an extreme environment. The questions on Worksheet 8 could be answered together on a large chart paper. A follow-up option is to share this learning with the large group.*

For enrichment, faster learners could create a 3-D model of the tool/instrument (in its modern day form) that they chose to research.

An Exploration Expedition!

Your teacher will be your guide as you go on an expedition to explore the extreme environment of Mars, the fourth planet from the sun, in our solar system.

Your task is to make and record observations of Mars. As the technology available to you advances, so should the discoveries you make about this extreme environment!

Let's Explore!

Stage 1:
Mars is, on average, about 225 000 000 km away from where you are standing on the planet Earth. Use **your eyes** to look at this object. Record your observations.

_____

_____

_____

Stage 2:
The year is now 1890, and you are using the latest in **binocular technology**. Use your binoculars to observe Mars. Record your observations.

_____

_____

_____

Stage 3:
The year is now 1930, and you are using the latest in **telescopic technology**. Use your telescope to observe Mars. Record your observations.

_____

_____

_____

OTM2157   ISBN: 9781770783638
© On The Mark Press

Stage 4:

The year is now 1960, and a space probe has been launched and has **entered the atmosphere** of Mars. Use the probe to gather data about Mars. Record your observations.

_____

_____

_____

_____

_____

Stage 5:

The year is now 1970, and another space probe has been launched and has successfully **flown by** Mars. Use the probe to gather more data about Mars. Record your observations.

_____

_____

_____

_____

_____

Stage 6:

The year is now 1980, and another space probe has been launched and has successfully **orbited** Mars. Use the probe to gather more data about Mars. Record your observations.

_____

_____

_____

_____

_____

OTM2157   ISBN: 9781770783638
© On The Mark Press

Name:

Stage 7:

The year is now 1990, and a space probe with enhanced camera action has been launched and has successfully orbited Mars. Use the probe to gather data about Mars.

Picture of Mars:

Stage 8:

The year is now 2005, and another space probe has been launched and has successfully landed on Mars. Use the probe to gather more data about Mars.

Description of surface:

_____

_____

_____

Description of air quality:

_____

_____

_____

OTM2157   ISBN: 9781770783638

| Name:

Stage 9:

The year is now 2015, and a space probe has been launched and has successfully **landed and is roving** on Mars. Gather data about Mars at different location points.

Detailed image of one side of Mars:

Caption:

Detailed image of the other side of Mars:

Caption:

OTM2157   ISBN: 9781770783638
© On The Mark Press

A Timeline of Technology

Revisit the list you made of the tools/ instruments that are used to access and explore the extreme environment that you researched in the previous lesson, titled "Adventuring into the Extreme!", Worksheets 3 – 6.

Choose one of the tools/ instruments on the list. Create a timeline of its technological advances, to show how its improvements have allowed more discoveries to be made in the extreme environment that you chose.

Let's Resarch It!

Name of tool/ instrument: _____

What can this tool/instrument do?

_____
_____
_____
_____

Explain the importance of this tool/ instrument to the exploration of the extreme environment that it is used in.

_____
_____
_____
_____
_____
_____
_____
_____

OTM2157 ISBN: 9781770783638

Name:

Years | Technological Advancements

Use the time line below to indicate your tool/instrument's technological advancements. Briefly describe each of its **major** advancements, and the years in which they were made.

Name:

Diagram of the tool/instrument in its early stages:

Diagram of it as it appears today (improvements are indicated):

OTM2157   ISBN: 9781770783638
© On The Mark Press

A Contribution from Home

Many advances have been made in technological devices that are used to access or to explore extreme environments.

Your task now is to research a technological advancement that has been made by **your country** as a contribution to the exploration of an extreme environment.

Let's Research it!

What is the technology?

_____

_____

Who invented it?

_____

_____

When was it first used?

_____

_____

Where has it been used?

_____

_____

_____

How has it contributed to the exploration of an extreme environment?

_____

_____

_____

_____

OTM2157   ISBN: 9781770783638
© On The Mark Press

ABORIGINAL CONTRIBUTIONS

LEARNING INTENTION:

Students will learn about the contributions of Aboriginal technologies to the exploration of extreme environments.

SUCCESS CRITERIA:

- research and compile information about an Aboriginal technology that has contributed to the exploration of an extreme environment
- record findings with the use of illustrations and written responses
- participate in a gallery walk to observe the work of classmates

MATERIALS NEEDED:

- a copy of "Examining an Aboriginal Innovation" Worksheet 1, 2, 3, and 4 for each student
- a copy of "A Gallery Walk of Innovation" Worksheet 5 and 6 for each student
- access to the internet
- Bristol board (2 pieces for each student)
- glue, scissors, pieces of card stock
- clipboards, pencils, pencil crayons, markers

PROCEDURE:

*This lesson can be done as a long lesson, or divided into two or three shorter lessons.

1. Discuss with students the term 'Aboriginal', to ensure their understanding. With access to the internet, students will research the contributions of Aboriginal people to the exploration of extreme environments. Give students Worksheets 1, 2, 3, and 4. On Worksheet 1, they will choose a category, which will guide their research. Finished work should be displayed on a Bristol board, which will be used for the next activity.

2. Give students Worksheets 5 and 6 on a clipboard. With the work from the previous activity displayed about the room, students will go on a gallery walk to observe the work compiled by their peers. They will choose four pieces of Aboriginal technology (that is different from what they had researched), to illustrate and describe its usage in extreme environment explorations.

DIFFERENTIATION:

Slower learners may benefit by working with a strong peer in order to 'navigate' the research required to complete Worksheets 1, 2, 3, and 4. Sections of the report could be divided between the partners to lessen the work load.

For enrichment, faster learners could design a board game that incorporates the Aboriginal technologies in order to explore extreme environments, with both positive events and obstacles occurring in order maximize the need for the technology.

OTM2157   ISBN: 9781770783638
© On The Mark Press

Examining an Aboriginal Innovation

Aboriginals have historically participated in and contributed technologies for the exploration of extreme environments. Their technological inventions have contributed to transportation forms used to access some extreme environments, shelters to weather environments, harvesting methods and tools used for food and survival, and clothing types to function comfortably in these environments.

Your task now is to research an Aboriginal technology that has been used in exploration. Check off the category that you will research:

⬜ **Transportation**　　⬜ **Shelter**

⬜ **Harvesting**　　⬜ **Clothing**

This is a diagram of _____

Name:

Illustrate and label the extreme environment(s) that this technology is used in.

OTM2157   ISBN: 9781770783638

List the materials that are used to create this **tech**nology.

- _____
- _____
- _____
- _____
- _____
- _____
- _____
- _____
- _____
- _____

Explain how this technology has contributed to the exploration of an extreme environment.

_____
_____
_____
_____
_____
_____
_____
_____
_____
_____
_____
_____
_____

OTM2157   ISBN: 9781770783638
© On The Mark Press

Name:

Years   **Technological Advancements**

Use the timeline below to indicate when advancements have been made in this technology. Briefly describe three of its **major** advancements.

OTM2157   ISBN: 9781770783638
© On The Mark Press

Name:

A Gallery Walk of Innovation

Your classmates have researched and gathered a lot of information about Aboriginal technologies that are used in the **exploration of extreme** environments. Let's take a gallery walk to **observe** and learn about these technologies!

Choose four technologies that are different **from** the one that you researched. Illustrate them, and describe **how they** are used in the exploration of an extreme environment.

OTM2157   ISBN: 9781770783638
© On The Mark Press

Continue your gallery walk. Illustrate and describe Aboriginal technologies used in the exploration of extreme environments.

OTM2157   ISBN: 9781770783638
© On The Mark Press